Water
109

老虎请喝茶

Tea for Tigers

Gunter Pauli

[比] 冈特·鲍利　著

[哥伦] 凯瑟琳娜·巴赫　绘

郭光普　译

上海远东出版社

丛书编委会

主　任：田成川

副主任：闫世东　林　玉

委　员：李原原　祝真旭　曾红鹰　靳增江　史国鹏

　　　　梁雅丽　孟小红　郑循如　陈　卫　任泽林

　　　　薛　梅　朱智翔　柳志清　冯　缨　齐晓江

　　　　朱习文　毕春萍　彭　勇

特别感谢以下热心人士对童书工作的支持：

匡志强　宋小华　解　东　厉　云　李　婧　庞英元

李　阳　梁婧婧　刘　丹　冯家宝　熊彩虹　罗淑怡

旷　婉　王靖雯　廖清州　王怡然　王　征　邵　杰

陈强林　陈　果　罗　佳　闫　艳　谢　露　张修博

陈梦竹　刘　灿　李　丹　郭　雯　戴　虹

目录

Contents

一群印度象正在阿萨姆河谷的河漫滩上漫步。这里是印度的东北部，母象们一直在警惕着周边的危险。附近有很多老虎，就连犀牛都不愿意让孩子离开自己的视线。

"这些年来，我一直以为人类是我们最大的问题。"母象说。

A herd of Indian elephants is roaming around the flood plains of the Assam River system. Here in the north-east of India, the elephant cows are always on the lookout for danger. There are so many tigers around, even the rhinos do not like to let their calves out of their sight.

"For years I thought that humans were our greatest problem," says the elephant cow.

一群印度象......

A herd of Indian elephants ...

"你说的一定是那些沿着河流建造村庄的人，他们能在每年河水泛滥造就的肥沃土地上耕种。"犀牛回答。

　　"当一些有远见的英国人决定把我们世代生存的土地建成保护动物的国家公园时，我们确实松了一口气。"

"You must be talking about those people who build their villages along the rivers so they can farm the rich soil created by the annual flooding of the river," responds the rhino.
"We did get some relief when some British visionaries decided to turn the land we need to feed ourselves and our young into a national park, a place reserved for animals only."

"这真是太有远见了。在这之前犀牛只剩不到100头了！几百年来，我们一直遭受偷猎，人们还侵占我们的土地。但是现在我们有世界上最大的野生犀牛种群了！"

"我们的数量也从不到100头增加到1 000多头。"大象回应道，"我们在这绿草地上茁壮成长。"

"That was an incredible show of foresight. Before that there were less than one hundred surviving rhinos! For hundreds of years we have suffered from poaching, and people encroaching on our land. Now we have the largest wild rhino population in the world."

"We have also expanded our numbers, from less than one hundred to more than one thousand," Elephant responds. "We thrive on these fields of green grass."

……最大的野生犀牛种群……

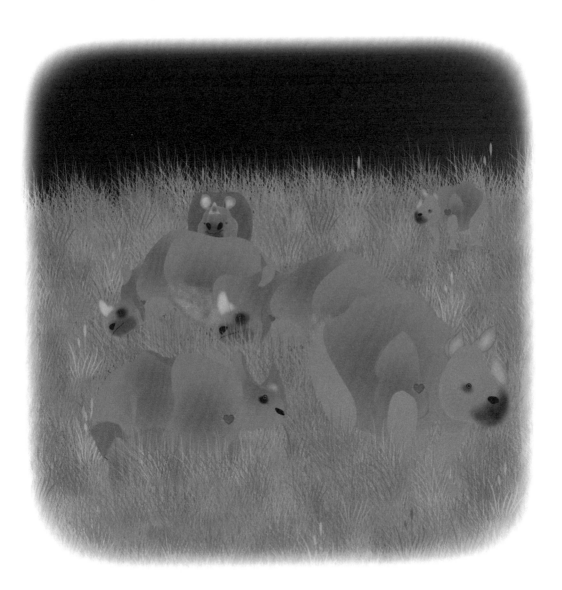

...the largest wild rhino population ...

......老虎的数量也变多了。

... there are also more tigers.

"所有生活在这个伊甸园里的动物都享受着大自然丰厚的馈赠。但是随着我们数量的增加，鹿和野猪也在增加，老虎的数量也变多了。单单这个公园里就有100多只老虎，我想你明白这意味着什么：我们的孩子不再安全了。"

"生活总是充满了危险，不过周围有100多只老虎意味着这里有足够满足他们的猎物。我们必须尽力保护我们的孩子，不让他们成为老虎的猎物。"

"Everyone living in this Garden of Eden enjoys the abundance offered by Nature. But as our numbers increase, and those of deer and wild pigs, there are also more tigers. This park alone has over one hundred and you know what that means: our little ones are not safe anymore."

"Life is always full of dangers, but having more than a hundred tigers around shows that there is enough prey for all of them. We have to do what we can to protect our little ones, so they do not fall prey to the tigers."

"这很难。"犀牛说，"你知道的，当平原被洪水淹没时，我们除了穿过人类的土地，没有别的选择。"

"是的，但至少茶园为我们提供了一条通往高地的廊道。"

"但在那里，老虎能轻易地攻击我们。"

"That is difficult," Rhino says. "You know that when the plains are flooded we have no option but to cross the people's land."

"True, but at least the tea plantations provide a corridor for us to find our way to highlands."

"But that is where the tigers will attack us with ease."

......茶园提供了一条廊道......

... tea plantations provide a corridor ...

……从前我们不得不踏过茶园……

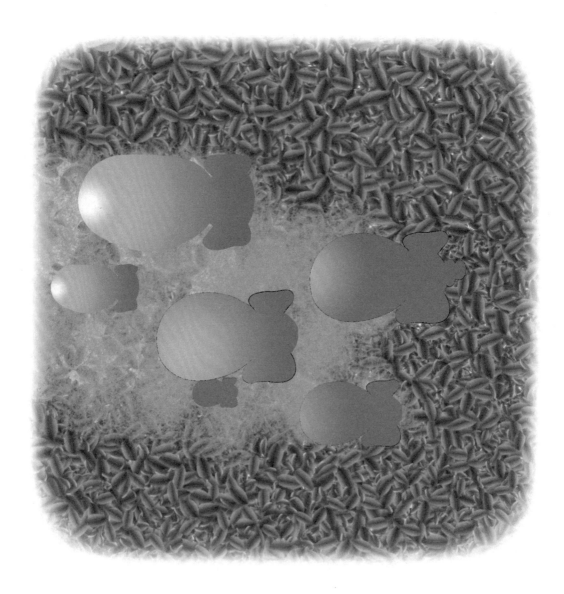

… before we had to trample through the tea fields …

"那你是宁愿淹死在河水里，还是冒险向高地转移？"大象问。

"哦，我们必须冒险。而且我们很感激这些廊道提供的安全路线。从前，我们不得不踏过茶园和人类的后院才能到达安全地点。"

"Now, would you rather drown in the river or take the risk and move to higher altitudes?" Elephant asks.

"Oh no, we have to take the risk. And we are grateful for this safe passage provided by the corridors, as before we had to trample through the tea fields and people's backyards to get to safety."

"茶农们很伟大。他们不仅给我们提供了通向安全地区的通道，甚至还停止使用有害的化学物质，以免我们受到毒害。"

　　"我知道。"犀牛说，接着她又补充道，"这对他们来说很不容易，停用那些化学物质让他们的茶叶产量受损。这里的土壤太贫瘠，茶树需要依赖那些化学物质才能长得更好。"

"The tea people have done a great job. They not only gave us a corridor to safety, they have even stopped using harmful chemicals so that we are not exposed to toxins."

"I know," Rhino says, and then she adds, "That could not have been easy for them as not using those chemicals has led to suffering a loss in production. The tea bushes relied on those chemicals to grow better, as the soil quality here is so poor."

······停止使用有害的化学物质······

...stopped using harmful chemicals ...

······贫瘠的土地!

... the poor soil!

"是的，茶树在富含碳元素的土地上生长得更好。"

"我听说这些茶是有机的，但是这个有机标签只能告诉我们这茶里没有添加什么物质。它完全没有提及贫瘠的土地！"

"你是说，假如种植者们没法在这里赚到钱，村民们就会丢掉自己的工作？这对我们来说真是坏消息，那样的话他们就会开始屠杀我们来得到我的象牙和你的犀牛角了。"

"Yes, tea grows well where there is a lot of carbon in the soil."

"I heard this tea is organic, but by giving it an organic label, it only tells us about what is not in the tea. It doesn't say anything about the poor soil!"

"Do you mean to say that when the growers do not make money here, the villagers will lose their jobs? That will be very bad news for us, because they will then start to kill us, me for my tushes and you for your horn."

"哦不！"犀牛大喊，"我们能做点儿什么？"

"我相信，那些善良的人能给我们提供通往山上的安全通道，那他们一定会有灵光一现的时刻，想出方法来确保所有人都有工作。这样我们都能够健康快乐地生活了。"

"我希望他们的灵光尽快出现——在他们还清醒的时候……"

……这仅仅是开始！……

"Oh no!" Rhino exclaims. "What can be done?"

"I am sure that those people who have the good heart to give us safe passage to the hills, will have some truly creative moments and imagine or dream up a way in which to secure that everyone here has a job. That way we will all be able to live healthy and happy lives."

"I hope that they have that dream soon – with their eyes wide open …"

... AND IT HAS ONLY JUST BEGUN! ...

······这仅仅是开始！······

...AND IT HAS ONLY JUST BEGUN! ...

Did You Know?

你知道吗?

There are currently around 2,400 one-horned rhinoceroses in the Kaziranga Park in north-east India. That is two-thirds of the world's population. When the Park was set up in 1905, there were only about a hundred left.

目前印度东北部卡齐兰加国家公园约有 2 400 头独角犀牛，占全世界犀牛种群数量的三分之二。1905 年公园刚成立时只有 100 头左右。

$6,000
100g

Poaching is a major threat to the survival of this unique animal. At US$ 6,000 for 100 grams, rhino horn is more expensive than gold. Over the past 25 years about 500 rhinos were poached by illegal horn traders.

偷猎是威胁犀牛这种独特动物生存的主要原因。100 克犀牛角售价 6 000 美元，比黄金的单价还要贵。在过去的 25 年里，非法犀牛角商人偷猎了约 500 头犀牛。

The Kaziranga Park has the highest density of tigers of any of the protected areas in the world. The tigers feed mainly on elephant calves, buffalo and swamp deer.

卡齐兰加国家公园的老虎密度是世界上所有保护区中最高的。这里的老虎主要捕食大象幼崽、水牛和泽鹿。

The Park was declared a protected area already in 1905. It houses 9 types of monkeys, 15 types of turtles, 42 types of fish and 2 types of pythons.

该公园在 1905 年被宣布成为保护区。这里有 9 种猴子、15 种龟、42 种鱼类和 2 种蟒蛇。

Large numbers of vultures die as a result of feeding on the carcasses of animals that are contaminated by anti-inflammatory drugs consumed by people, which have ended up in the food chain.

人类使用的消炎药会在食物链中留存，大量秃鹫由于食用了消炎药污染的动物尸体而死亡。

The largest area of the Park is covered in high grassland, which offers an ideal habitat for tigers to hunt in.

卡齐兰加国家公园中占地面积最大的是高草地，这为老虎捕猎提供了良好的环境。

When the heavy monsoon rains start, all the animals in the Kaziranga Park need to leave the plains in search of safe ground in the hills. The Government built artificial highlands for shelter and tea plantations offer corridors for safe passage.

当暴雨季开始时，卡齐兰加国家公园所有的动物都得离开平原，去山上寻找安全的地方。政府修建了人工高地庇护所，而茶园则为动物们安全通过提供了通道。

When neighbouring Hathikuli Tea Estate stopped using chemicals, water pollution from the run-off of agro-toxins ceased. However, the drop in output forced the tea estate to diversify into producing mushrooms and black pepper to ensure full employment in the region.

当附近的赫蒂库利茶场停止使用化学品后，就不会再有农业毒素通过地表径流污染水资源了。然而，产量的下降迫使茶园转向多元化生产，同时种植蘑菇和黑胡椒，以确保该地区的就业。

Think About It

想一想

你愿意保护濒危的动物吗？你愿意为野生动物提供安全的区域，使它们能够在自然环境中繁衍生息吗？

Would you like to protect animals that are endangered, or would you like to provide safe areas where wildlife can regenerate to its natural carrying capacity?

你会为了钱而猎杀濒危动物吗？

Would you kill animals that are critically endangered, for money?

如果一个人没有工作并且失业多年，他怎么能保证子女的未来？如果你是一个没有工作的人，你会为了钱而猎杀动物吗？

How can people, who do not have a job, and have not had one for years, ensure that their children have a future? Would you kill an animal for money if you were some one without a job?

老虎捕猎大象和犀牛的幼崽，你会去保护这些幼崽吗？

Tigers hunt for elephant and rhino calves. Would you protect the calves?

Do It Yourself!

自己动手！

Is there a nature reserve close to where you live? If so, find out how much wildlife there is. How easy is it for people to live next to a place where wild animals live? Is there enough space for animals, plants and birds to live, or are more and more people coming to live in the area, attracted by the park? Is the park slowly but steadily becoming smaller and smaller? What measures would you propose to ensure that Nature and people live in harmony? Develop a plan, and propose this to your friends. Be sure to be ambitious.

你住的地方附近有自然保护区吗？如果有，找出那里有多少野生动物。人们在野生动物居住地附近生活容易吗？那里有足够的空间让动植物生存吗？是否有越来越多的人被公园所吸引而来这里居住？这个公园有没有持续在慢慢缩小？为了人和自然能和谐相处，你有什么建议？制订一个计划并讲给你的朋友听。一定要充满雄心壮志。

学科知识
Academic Knowledge

生物学	野生动物保护区在保护生物多样性中的作用；每年的洪水对于生态系统的重要性；亚洲象和非洲象之间的差异；独角犀牛和双角犀牛之间的差异；非洲五大动物：大象，犀牛，水牛，狮子，豹子；印度五大动物：亚洲大象，独角犀牛，水牛，老虎和泽鹿。
化 学	活性农药、除草剂和杀菌剂的功能和其在土壤中积累以及流失到水体中带来的长期影响没有被联系起来；高浓度有机质在土壤中十分重要，它能使土壤肥沃。
物 理	水平面上升的物理作用对冲积平原很重要，在净化该地区的同时还能补充养分。
工程学	使用无人机监视某些地区并防止偷猎。
经济学	保障充分就业是消除偷猎的前提条件；野生动物栖息地的生态系统服务功能；基因库和固碳的价值。
伦理学	为什么有些人会为了农业和医药而排放永久性毒素，即使这会影响动植物生存并破坏表层土壤和所有人的生活？
历 史	1757年到1858年殖民者通过东印度公司控制印度；1858年到1947年英国对印度进行殖民统治，将其纳入大英帝国；公元前300年，天爱帝须王宣布斯里兰卡的米欣特莱周边地区成为野生动物保护区，这是历史上第一个野生动物保护区。
地 理	东洋生态区；雅鲁藏布大峡谷；泰莱-杜阿大草原；生物多样性热点；印度次大陆一般只有三个季节：夏季、雨季、冬季。
数 学	定性数据和绝对数字的结合；生态系统通过物理、化学和生物过程自我维持稳态，生态系统的功能是价值中性的，但这些功能所产生的结果对社会是有益的。
生活方式	人们参与野生动物保护的一般意愿；我们的目的是保护某些物种免遭灭绝，还是让自然回归到进化道路上来产生生物多样性？很少有人愿意为生态友好产品花费更多；我们接受"有机"这个标签，即使它只说明食品中"不"含有什么。
社会学	人们愿意在水源周围定居。
心理学	不愿意冒险的母亲对保护后代准备更充分；面临必然来临的死亡和不愿冒险之间的两难抉择；白日梦与深度睡眠时所做的梦的对比。
系统论	国家公园的重要目的是保护生物多样性，而茶场面临着生产更低价茶叶的压力，只能以较低的成本来增加产量，这二者之间需要平衡。

情感智慧
Emotional Intelligence

母　象

大象在开始时承认她原来的错误认识。她也承认一些人类在野生动物保护方面是很有远见的。她很感激人们提供的保护和生活条件，让濒临灭绝的种群数量得以恢复。她认为：生活中存在风险，所以父母需要采取必要的、额外的预防措施。大象质疑了犀牛对于通过廊道到达安全地区的看法，她认为：要么冒险，要么死亡。大象很感激那些停止使用有害化学物质的人们，也对这样影响了人类的生活表示惊讶。因为严重的失业状况，贫穷的人将会去帮助偷猎者，这将使动物处于危险中。大象希望当地的人们找到新的生活目标，这样可以让动物的安全有保障。

独角犀牛

犀牛很认可大象的观点，承认了人类建立野生动物保护区的远见，还为此提供了证据。犀牛觉得在这里有天堂般的生活，享受着充足的食物和生物多样性，但她意识到了这里存在着危险。犀牛很感谢穿过茶园通往高地的廊道，因为这让他们避免了践踏人们生活区的危险。犀牛了解人们面对的损失。她观察到"有机"这个标签只是说明了茶中不含有某些成分。犀牛敦促人们思考超越现有状态的限制，努力做到充分就业，这样自然中的生灵和人类都可以幸福健康地生活。

艺术
The Arts

这一次，你需要面临的挑战是画三幅画：分别描绘雅鲁藏布江的夏季、冬季和雨季。通过比较，我们可以看到它在三个季节里的基本变化，特别是在雨季。这有助于我们了解雨季意味着什么，以及它如何改变了生活。不需要描绘得非常仔细，但是要用不同的颜色表达出显著的变化。

思维拓展
Systems: Making the Connections

随着人口数量的增加，越来越多的人入侵到野生动物生存的土地，这很大程度上导致了野生动物的数量下降——某些物种已经濒临灭绝。我们正在目睹几千年来野生动物最大的一次消亡，而且这是第一次由人类造成的。建立国家公园和野生动物保护区是富有远见的创举。目前，对抗人类活动扩张而保护动植物的行为只取得了有限的成效。当贫穷越来越严重，而周围都是自由活动的动物时，人们会为了生计而猎杀，这会导致更多动物的死亡。即使是运用强制力量保护野生动物，政府也遇到了巨大困难。严重的饥饿使人们为了生存而入侵这些土地进行捕猎和耕种。要成功管理一个保护区，需要在公园内及周边发展新型经济。一项政策要旨在满足所有基本需求：水、食物、住房、卫生、能源和教育。只有所有人都有工作，并且报酬高于最低薪资，社会和医疗服务得到保障，人们才不会出于对食物和利益的追求去捕猎。不幸的是，这些地区的保护采取的是压制策略，派驻配备高科技装备的警察，但是依然有偷猎者愿意冒着生命危险去猎杀稀有动物来赚一大笔钱。世界上几乎所有的国家公园都有主要物种数量缩减的现象。直到最近，这个现象才好转。卡齐兰加国家公园是一个例外，该公园的核心物种——犀牛、大象和老虎等的数量都有实际增加。此外，茶场经历了重大转型，从传统的殖民种植和加工产业演变为利用当地资源促进地区发展的多样化经济，把这个地区最好的产品呈现给世界。

动手能力
Capacity to Implement

稍微想象一下茶园能用什么方法拯救老虎和犀牛。一个年产量只有300吨，在野生动物保护区旁的有机绿色茶场，是怎样成为经济发展的引擎和平台，甚至提供全民就业的？想一想，老虎和茶的关系是什么？还有什么种类的生物可以在茶场内和附近繁荣发展并带来更多收入？

故事灵感来自
This Fable Is Inspired by

兰吉特·伯尔特库尔
Ranjit Barthakur

兰吉特·伯尔特库尔出生并成长于印度阿萨姆邦。他提出了自然经济的概念，促进了基于农业和生物多样性的社区行动。他还提出了自然第一（Nature First）这个概念，目的是通过土地、能源、废物、水、空气和碳等方面来恢复自然之间的平衡。塔塔咨询公司也向他的管理和发展概念公司学习。他在大象身上得到了很大启发，作为综合茶场公司的董事长（与卡齐兰加国家公园接壤的赫蒂库利有机茶场就属该公司下辖），他启发了股东，与当地居民找到策略，实现有机茶叶带来的经济发展和保护野生动物之间的和谐，从而确保当地社区的文化、传统和智慧得到加强。

图书在版编目(CIP)数据

冈特生态童书.第四辑:修订版:全36册:汉英对照 /
(比)冈特·鲍利著;(哥伦)凯瑟琳娜·巴赫绘;
何家振等译.—上海:上海远东出版社,2023
书名原文:Gunter's Fables
ISBN 978-7-5476-1931-5

Ⅰ.①冈… Ⅱ.①冈…②凯…③何… Ⅲ.①生态环
境−环境保护−儿童读物—汉、英 Ⅳ.①X171.1−49

中国国家版本馆CIP数据核字(2023)第120983号
著作权合同登记号图字09-2023-0612号

策　　划　张　蓉
责任编辑　张君钦
封面设计　魏　来李　廉

冈特生态童书
老虎请喝茶
[比]冈特·鲍利　著
[哥伦]凯瑟琳娜·巴赫　绘

郭光普　　译